Seymour Sim



# FIGHTING FIRES

SEASTAR BOOKS • NEW YORK

**This book is dedicated to Robert and Nicole.**

Special thanks to reading consultant Dr. Linda B. Gambrell, Director, School of Education, Clemson University. Dr. Gambrell has served as President of the National Reading Conference and Board Member of the International Reading Association.

Permission to use the following photographs is gratefully acknowledged: front cover, pages 6–7: © Mark L. Stephenson/CORBIS; title page: © Peter Skinner, Photo Researchers, Inc.; pages 2–3: © Gary Braasch/CORBIS; pages 4–5: © Paul A. Souders/CORBIS; pages 8–9, 12–23, 26–27, 32: © George Hall/CORBIS; pages 10–11: © Gunter Marx Photography/CORBIS; pages 24–25: © David Rowley, New England Stock; pages 28–29: © Michael S. Yamashita/CORBIS; pages 30–31: © Peter B. Kaplan, Photo Researchers, Inc.

 First published in the United States by SeaStar Books, a division of North-South Books Inc., New York. Published simultaneously in Canada, Australia, and New Zealand by North-South Books, an imprint of Nord-Süd Verlag AG, Gossau Zürich, Switzerland.

Library of Congress Cataloging-in-Publication Data is available.
ISBN 1-58717-168-6 (reinforced trade edition)
1 3 5 7 9 RTE 10 8 6 4 2
ISBN 1-58717-169-4 (paperback edition)
1 3 5 7 9 PB 10 8 6 4 2
PRINTED IN SINGAPORE BY TIEN WAH PRESS
For more information about our books, and the authors and artists who create them, visit our web site: www.northsouth.com

Smoke jumpers are firefighters who parachute from airplanes. Their work is very dangerous. They have no way to escape if the wind shifts and the wildfire burns toward them.

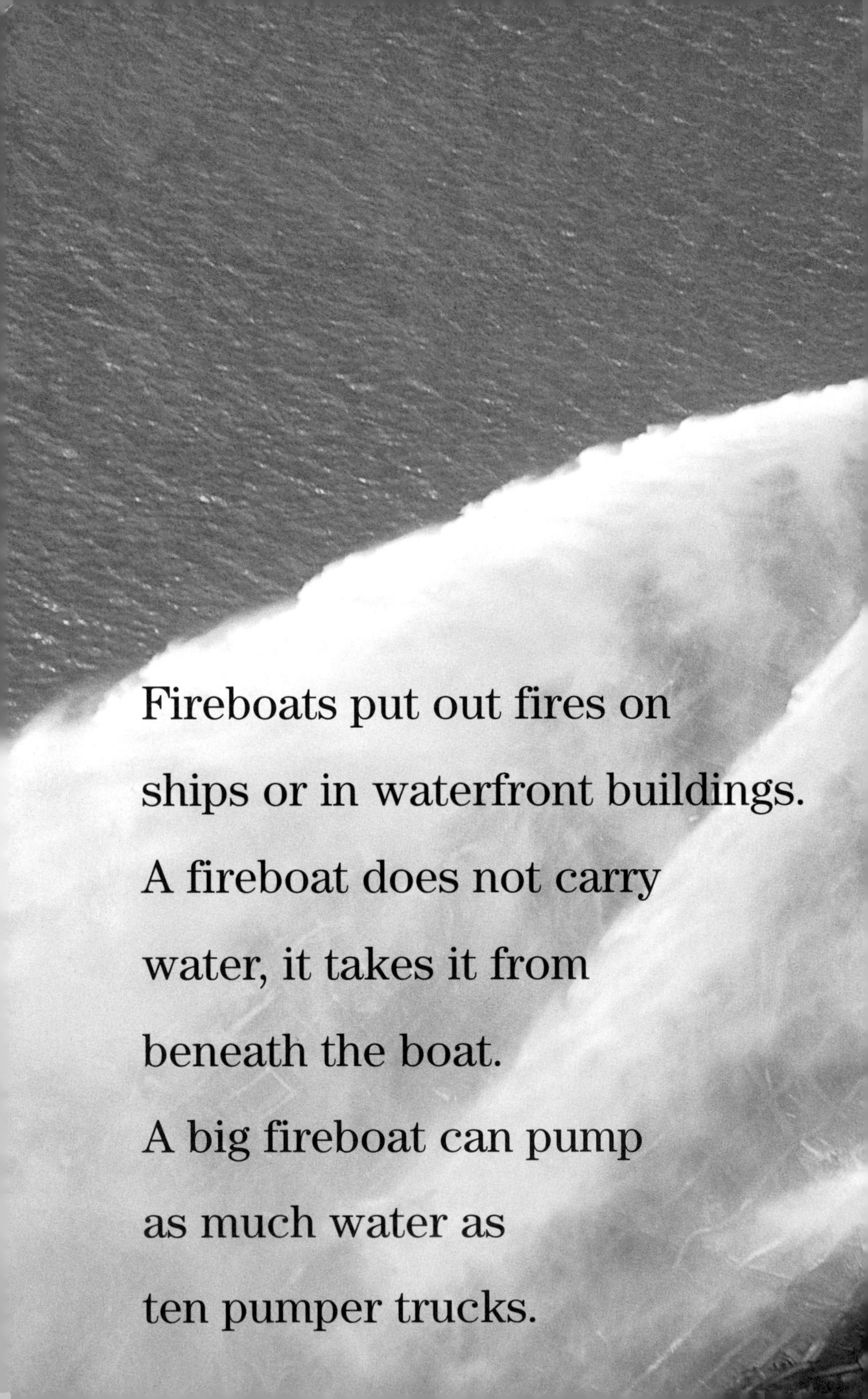

Fireboats put out fires on ships or in waterfront buildings. A fireboat does not carry water, it takes it from beneath the boat. A big fireboat can pump as much water as ten pumper trucks.

If you liked learning about **Fighting Fires**, you will also enjoy

## LEVEL 1

**Baby Animals**

**Giant Machines**

**Killer Whales**

**Planets Around the Sun**

**Wild Bears**

## LEVEL 2

**Amazing Aircraft**

**Danger! Earthquakes**

**Danger! Volcanoes**

**Super Storms**

***SEYMOUR SIMON*** is a former teacher and the author of more than 200 science books for children, more than half of which have been named Outstanding Science Trade Books for Children by the National Science Teachers Association. In addition to his series about the human body, space, and natural phenomena, Mr. Simon has written numerous books about earth's most compelling creatures, including *Whales*, *Gorillas*, *Sharks*, *Wolves*, and *Crocodiles and Alligators*. He is also the author of *Animals Nobody Loves* and *Out of Sight: Pictures of Hidden Worlds*.

Mr. Simon is the recipient of many awards honoring the body of his work, among them the *Washington Post*/Children's Book Guild Award; the Hope S. Dean Memorial Award, presented by the Boston Public Library; and the Knickerbocker Award for Juvenile Literature, presented by the New York Public Library Association.

Mr. Simon lives in Great Neck, New York. To see more, visit his web site at www.seymoursimon.com.

Front cover photograph © Mark L. Stephenson/CORBIS
Back cover photograph © Gunter Marx Photography/CORBIS
Designed by Matthew Siee

ISBN 1-58717-169-4
UPC 00169
6 51495 00395 5

$3.95
Higher in Canada

**In just one minute, pumper fire trucks can shoot out 1,250 gallons of water. That's enough water to fill 50 bathtubs.**

**Race into**

# FIGHTING FIRES

**and *SeeMore!***

INCLUDES 4 COLLECTIBLE CARDS

SeaStar Books
NEW YORK
PRINTED IN SINGAPORE

ISBN 1-58717-169-4
90
9 781587 171697

Special trucks, boats, and aircraft help us fight fires.

1170
STEIN
Breathing Air
2216
MSA

Firefighters risk their lives to put out fires and save other people's lives.

FIX URE CO.
9

Long ago, people passed buckets of water by hand to put out fires. Now, when an alarm sounds, firefighters have many kinds of special trucks to help them.

Pumper trucks carry water, hoses, and a pump. In just one minute, pumpers can shoot out 1,250 gallons of water.

That's enough water to fill 50 bathtubs. Many pumper trucks also make bubbly foam that puts out oil fires.

Some pumper trucks carry
a long crane called a boom.
Firefighters stand inside
a large bucket and are raised
100 feet into the air
to pump water or foam
on a fire.

Boston

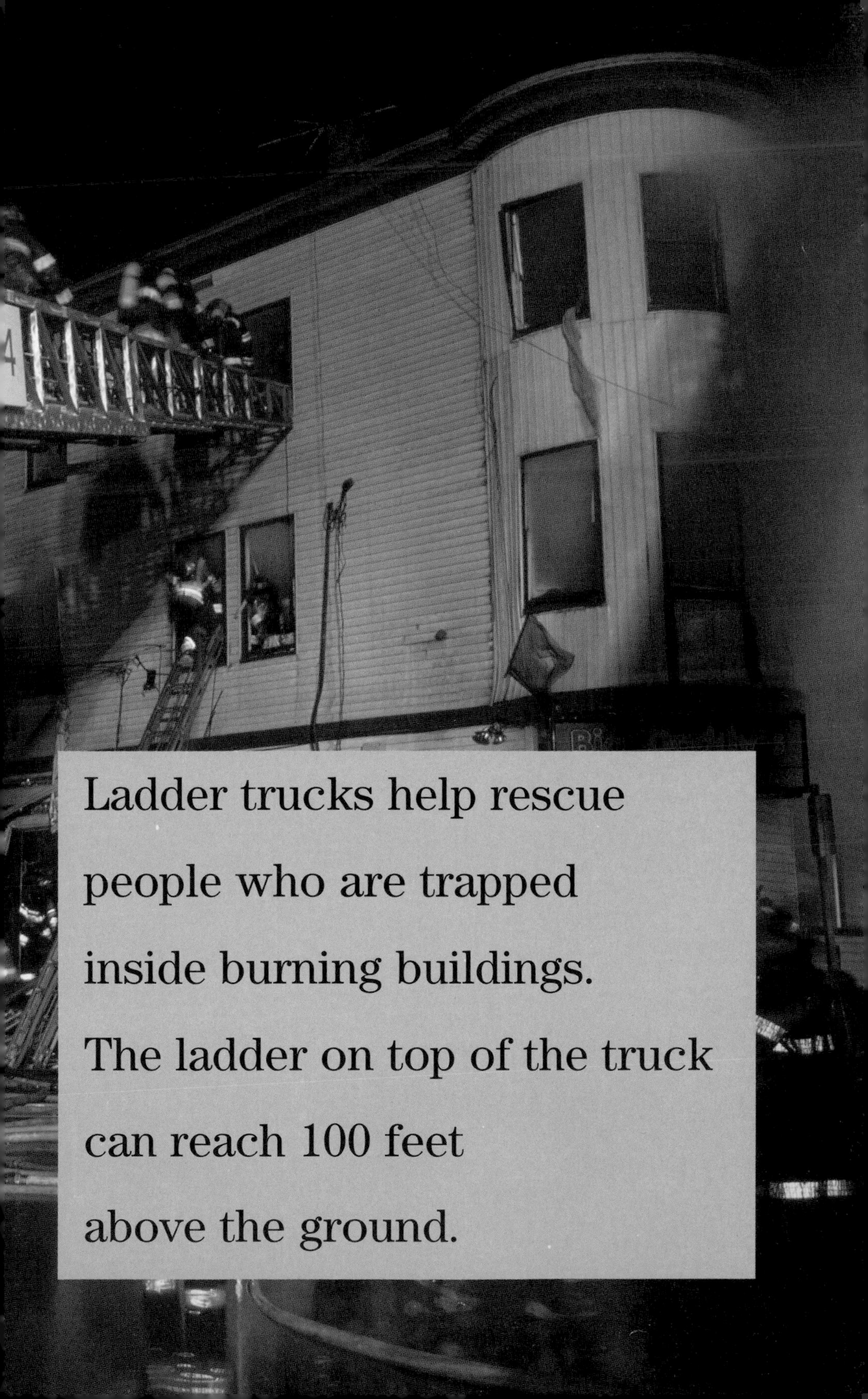

Ladder trucks help rescue people who are trapped inside burning buildings. The ladder on top of the truck can reach 100 feet above the ground.

SAN FRAN
T1

Tiller trucks are two ladder trucks that are joined together. One driver steers the front and a tiller driver steers the back.

Two drivers make it easier for the big truck to get through narrow city streets.

Quint trucks do the work of both pumper and ladder trucks.
A quint truck has a pump, a water tank, hoses, and two kinds of ladders.

Searchlight trucks are used to light up dark smoky fires so firefighters can see. The trucks' eight spotlights are bright enough to light a sports stadium at night.

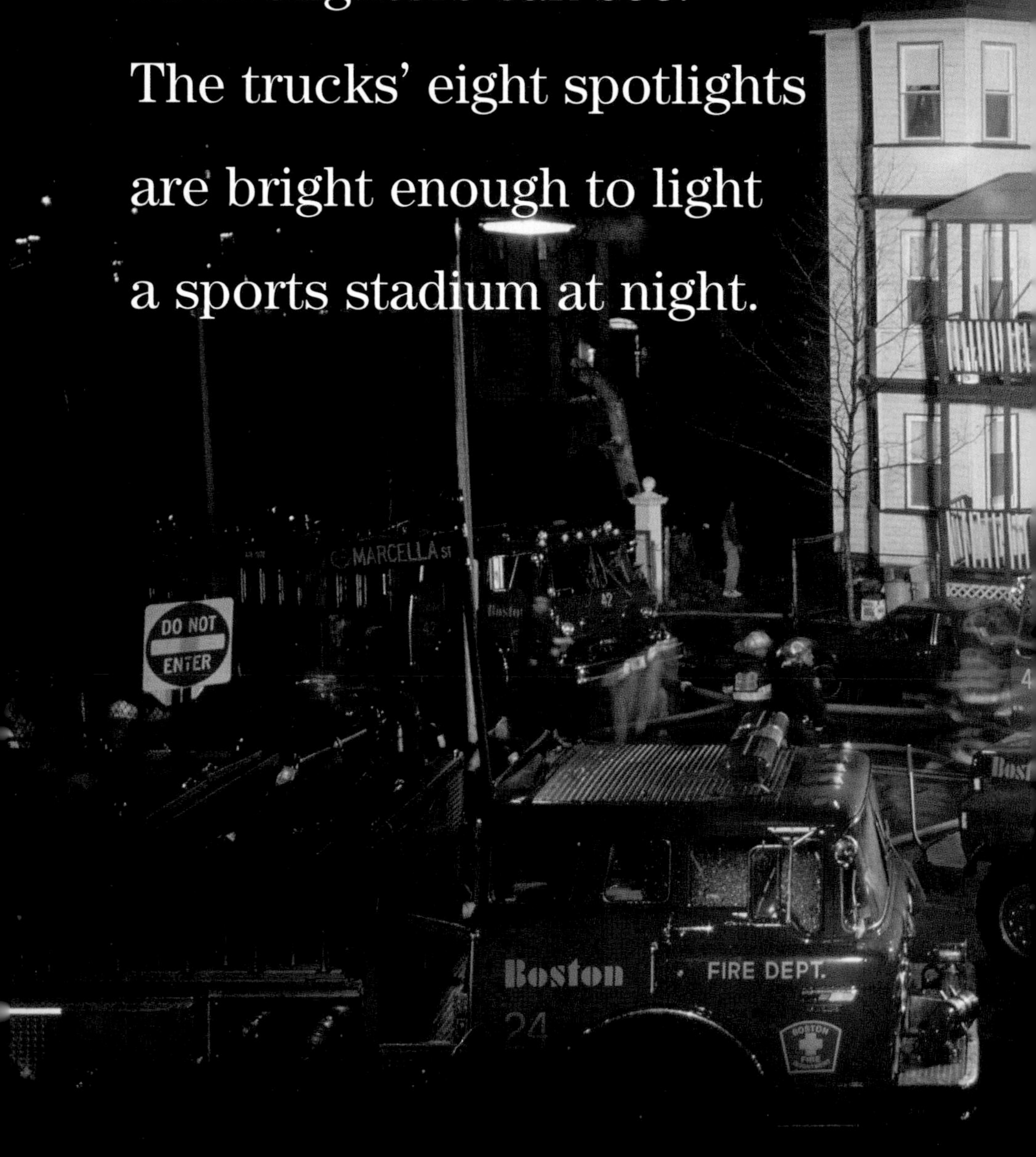

Boston LADDER 4
H
H2

Rescue trucks carry tools for emergencies, including special suits for walking through fire.

Rescue trucks also use the "Jaws of Life" to help remove people from crushed cars or collapsed buildings.

Aircraft crash rescue vehicles carry as much as 2,500 gallons of water.

That's because an airplane crash
may be a long way
from a water supply.
A crash rescue vehicle can put
out an airplane fire in minutes.

Off-road fire trucks
are used to fight forest fires
and brush fires.
They carry water, foam,
hoses, and pumps.

Off-road fire trucks need
to be tough enough
to drive in places
where there are no roads.

Special helicopters drop water
from 100-gallon buckets
onto forest fires.
A huge bucket scoops up water
from a lake for the next flight
over the flames.

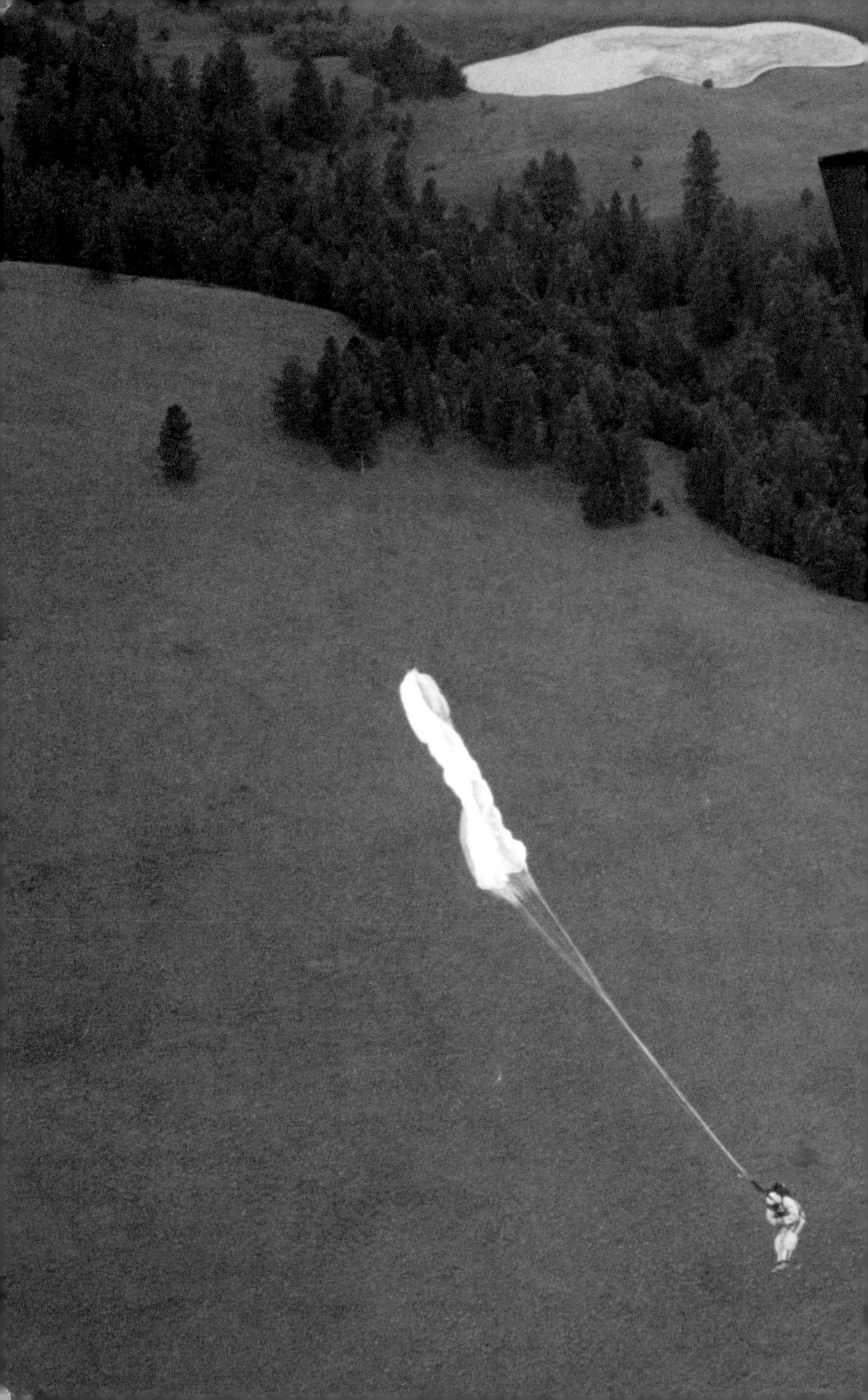